photo: Steve Schubert
Peregrine falcon in flight at Morro Rock

A peregrine falcon soars high above the 581-foot summit of Morro Rock. Gazing far in the distance with piercing eyesight, the falcon searches for avian prey below. Its keen vision is enhanced by dark malar stripes below the eyes that reduce the glare of the sun, and adaptation mimicked by athletes who paint black beneath their eyes. After a few rapid wing beats, the narrow, pointed winds are folded alongside the body and the streamlined predator plummets downward at incredible speeds that can exceed 200 miles-per-hour, the fastest animal on the planet. The momentum and force of this diving 'stoop' causes the prey to seemingly explode in a puff of feathers, injured or delivered a lethal impact by the strike of the feet and sharp talons. The falling bird is pursued and retrieved falling through the air or picked up from the ground, and if still struggling, quickly dispatched with a bite to the neck with a notch on the upper beak (the tomial or 'falcon's tooth'), severing the spinal cord. The falcon carries its captured prey often to a favorite feeding perch high up on the rocky face of Morro Rock, where feathers are plucked and drift away on the seabreeze and the sharp hooked beak tears away morsels of flesh to be consumed. These are the behaviors and adaptations of a highly efficient aerial predator, a hunter of the sky that catches it prey on the wing with astounding speed and agility.

The peregrine hunts a variety of available prey items including migrating shorebirds such as sandpipers and phalaropes, small ducks, occasional gulls, and terrestrial birds such as mourning doves, pigeons, swallows and swifts. The falcons hunt for avian prey in a diversity of terrestrial and aquatic habitats: above the steep shrub-covered slopes of Morro Rock, the nearby sandy beaches and dunes, over the nearshore waters of the Pacific Ocean, the Morro Bay estuary with its associated intertidal mudflats and salt marsh, and across the inland valleys and mountainous slopes of the Chorro Creek and Los Osos Creek drainages, which comprises the Morro Bay watershed.

During the nesting season the eggs are laid in a shallow depression, or "scrape", that was scratched out of the soft substrate on the floor of a cave, pothole, or rocky ledge, up high on a steep rocky escarpment. The eggs are less likely to roll out over the edge on these flat-bottomed nest sites. The falcon's nest is also referred to as the "eyrie". The location of nest sites on remote, steep cliffs protects the young from most terrestrial predators, such as raccoons, coyotes, and feral cats. The eggs, averaging 3 to 4 in a clutch, are incubated by both adults, hatching in about 33 days. The adult male hunts and brings back bird prey to the nest site, vocalizing with a food exchange call and transferring the prey in mid-air to the female - from his beak or feet to her gripping feet - as she maneuvers and flips agilely beneath her mate, a feat of superb flying skill and timing. The adult female, typical for most raptors, is about 1/3 larger in size than the male. The female takes the prey to the nest site, located several hundred feet above sea level on the steep cliff face of the Rock. She feeds the downy chicks - called "eyasses" - by tearing small morsels of flesh with her beak and tending to each hungry youngster. After the falcon chicks are several weeks old and growing rapidly, both parents will hunt prey - sometimes in tandem chases - and the parent will drop the prey inside the eyrie for the chicks to feed upon, often with considerable sibling rivalry occurring over the sudden appearance of food in the nest.

The nestlings take their first perilous flight and fledge from the nest at about 40 - 42 days of age. These fledglings will have dispersed and are usually gone from the vicinity of the Rock by late summer, although the resident adult falcons are present on territory and reside at Morro Rock year-round. Peregrine falcons usually mate for life. After the death or disappearance of one of the adults of a breeding pair, the remaining falcon may form a new pair bond – a mate replacement - with an individual from the 'floater' population, an unmated peregrine not occupying a nesting territory until a vacancy occurs.

Tiercel carrying shorebird prey (a Willet)

Female falcon in pursuit

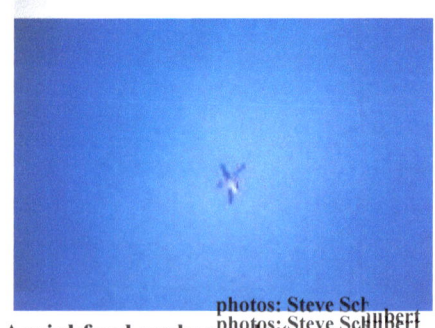
Aerial food exchange between the pair

photos: Steve Schubert

Peregrine falcons have likely nested at Morro Rock for many centuries. The territorial falcons fiercely chase away and attack other intruding peregrines and potential predators of their young, including hawks, eagles, and owls. During the nesting season the falcons execute surprise high-speed attacks and with their feet 'thump' turkey vultures, gulls, cormorants, and pelicans that fly too close to the eyrie. Human intruders illegally climbing around on top of Morro Rock cause an agitated, alarmed response with loud 'cak-ing' vocalizations and close flybys that can result in a strike from the falcon's sharp talons.

Peregrine attacking a Turkey Vulture at Morro Rock

The Anatum or American Peregrine Falcon, *Falco peregrinus anatum*, was designated a federally listed endangered species in the early 1970s. At Morro Rock there have been incidents of illegal shooting, theft of young falcons from the nests for the sport of falconry, and other human disturbances near the nest sites, including illegal climbing, close aircraft approaches, and rock quarrying blasting activities. Although concerns have been raised at specific nesting territories such as at Morro Rock where these kinds of activities have taken place in the past, none of these factors were likely responsible for the widespread and dramatic population decline of the species throughout much of its geographical range.

After years of research, it is now known the nearly worldwide decline of peregrine falcon populations following World War II was due to the introduction and widespread use of a organochlorine pesticide, DDT. DDE - the metabolic breakdown product of DDT - interferes with calcium metabolism and the formation of eggshells within the female's body. DDE causes the formation of exceedingly thin eggshells that break under the weight of the incubating parent. Loss of moisture through the thin shells can also dehydrate and kill the developing embryos inside. Dioxins and PCBs are other toxins that also biologically magnifies through ecological food chains and have been found as chemical residues within the falcons' body tissues and their developing eggs. The inability to successfully hatch eggs and reproduce is now known to have been the major cause of the peregrine falcon's dramatic population decline and imperiled status. Losses and disturbances of nesting and foraging habitats, in particular due to coastal residential and urban development and widespread loss of wetland habitats, and changes in prey availability may have also affected peregrine falcon populations on a local basis.

Morro Rock rises precipitously near the entrance of the Morro Bay harbor mouth and has served as an important navigational landmark for several centuries. The explorer Juan Rodriguez Cabrillo, approaching by ship in 1542, was the first European to observe and name Morro Rock. Located along the Central Coast of California, Morro Rock is an eroded volcanic neck or plug, which lies in a line with other volcanic peaks between the cities of Morro Bay and San Luis Obispo, locally referred to as the "Sisters". Hot magma cooled inside the throat of a now extinct volcano, forming a type of igneous rock named porphyritic dacite, dated at about 23 million years old.

The Pacific tides once ebbed and flowed completely around Morro Rock, which was then a nearshore island, but this isolation ended with the construction of the present causeway, creating an artificial land bridge connecting the Rock to the mainland. The causeway was built by the WPA (Works Project Administration) during 1933 to 1935. It provided access for pedestrians and vehicles, and closed the north channel by the Rock, creating a more sheltered harbor. However, this land bridge connection to the Rock along Coleman Drive has also at times promoted easy access, illegal climbing, and disturbances at what today is an Ecological Reserve and habitat for an endangered species and other wildlife.

Quarrying of rock for breakwater construction during the 1890s and the earlier part of the 1900's resulted in destruction of nearly one-third of Morro Rock's exposure. This intermittent blasting activity, which may have had a negative impact on nesting attempts by the peregrines, did not completely cease until after 1969.

Morro Rock lies within state park property and is an Ecological Reserve. Morro Bay has been designated a Globally Important Bird Area and is also registered in the National Estuary Program.

photo: SCPBRG files

Morro Rock aerial view

The Morro Rock location is among the most well-known peregrine falcon nesting sites in all of North America. Local residents, visitors, and tourists from around the world have learned about and observed the falcons here, often for the first time beholding with awe the sight of the falcons in flight and observing their nesting activities at this striking geographical landmark, sometimes simply called the 'Rock'. For many centuries, Native Americans residing in village sites near the Morro Bay estuary must have been aware of the peregrine falcons residing and nesting at Morro Rock, then a near shore island.

The Morro Rock site has received wide media publicity and played an important role locally in an endangered species conservation and management program, which, after several decades, assisted in the recovery of the peregrine falcon population throughout much of North America. The following synopsis discusses yearly nesting events, observations, and the management efforts at Morro Rock during the past several decades - a 50-year history. These are the stories of nesting successes and losses from year to year, innovative captive breeding and reintroduction techniques, dedicated falcon observers, and cautious renewed optimism for the recovery of a species once near extinction.

1967 Dr. Monte Kirven (San Diego Museum of Natural History) was informed by a falconer of an active nest on the east side of Morro Rock. Dr. Kirven made a request to the Morro Coast Audubon Society (MCAS), a local chapter of the National Audubon Society, to establish a volunteer nest guard duty. Vernon and May Davies volunteered for the nest watch, along with other MCAS members. Twenty volunteers in shifts observed from the parking lot below. Climbers were turned away several times during the season. MCAS succeeded in getting a State Park ordinance passed that prohibited climbing of Morro Rock.

One of two nestlings survived and fledged from the nest. Two versions exist for the fate of the other nestling, having either died in the nest or possibly having been stolen by falconers. It was not necessarily illegal to take peregrine young for falconry during this time, before peregrine falcons were given protected status as an endangered species. Note: it is probable that the Morro Rock eyries during the earlier 1960's and previous decades were robbed more than once of their nestlings, before the nest guard and management program was initiated. Falconers, egg collectors, and some biologists were certainly aware of the existence of peregrine falcons nesting at the Rock since the early decades of the century. Although there may have been attempts at egg collecting - a popular pastime earlier in the 20th century - there are no known private or museum collections of Morro Rock eggs still in existence.

1968 The Davies resumed the nest watch. Dr. Robert Risebrough (U.C. Berkeley) suggested that they observe the nest with binoculars and keep detailed notes, since opportunities to observe peregrine sites were now so rare. The peregrine falcon in North America had become extinct as a breeding species east of the Mississippi River, and was now rare throughout the West.

Three young hatched in June. One died after fledging.

Dr. Robert Risebrough and Dr. Monte Kirven climbed into the nest after the young had fledged. They examined the prey remains in the nest and found a large proportion of land birds, including mourning dove bones, suggesting that the falcons had been hunting inland.

1969 The Davies took on the nest watch full time, rather than scheduling and recruiting other volunteers. During the egg-laying period a fierce storm arrived. The Davies returned home, but were unable to return to the Rock the next day when the causeway was closed. Contractors blasted at a rock quarry below the nest day and night for rock rubble the city needed to repair a road washout. Detonations continued - the nesting cliff was floodlighted at night to allow blasting activities - until the contractors took down the barricades after 10 days. The disturbed falcons abandoned the nest and moved to a new site.

A few days later the adult female behaved strangely. She apparently would not eat. Vernon Davies found the adult female lying dead on a ledge Easter morning, the body sprawled on the Rock and feathers ruffling in the wind. Dr. Robert Risebrough and Dr. Steven Herman were called, making the long drive and arrived at 3:30 a.m. in the morning. They found the fallen dead falcon at dawn. A laboratory autopsy determined there was a prolapsed oviduct.

No young were produced this year.

1970 Dr. Steven Herman, a U.C. Davis biologist, searched more than 100 historic nesting sites in California and found 10 peregrine individuals, including only two known recently successful nesting sites in the entire state, one of which was at Morro Rock.

The tiercel (male falcon) appeared during the spring and remained alone all summer. No young were produced this year.

1971 A pair arrived and nested in the spring. It is not known if this was the same male with a new mate, or a new pair. The Davies bought and lived in a camper in the parking lot at the base of the Rock during the week, returning home only on weekends. Brian Walton, Carl Thelander, and Patrick "Nolan" Veesart maintained the watch on weekends, guarding the nest site day and night. The Davies frustrated nest robbing attempts by falconers…if they saw ropes or climbing gear in parked cars, they asked the owners about their intentions. Some went away after arrest threats were made, although others were abusive with curses and threats.

There were three young birds by late May. The first young fledged on June 6th, fell into a crevice and disappeared, but appeared the next day and was fed by the parents. The other two fledged during the next few days. The fledglings' flying skills were awkward at first, but developed quickly and they were soon taking food from their parents in flight.

1972 Nesting began by mid-March. After being seen on top of the Rock May 8th, one man descending was arrested by Fish and Game warden Howard Martin. He claimed he was looking for another falconer (who was taking the young while hidden by the heavy fog). The next morning no feeding of the young caused the observers to be alarmed. Biologist Ron Garrett climbed to check the nest. The two three-week-old nestlings were missing. A climber's homemade climbing aid, cigarette butts, and candy wrappers were found on top of the Rock.

John Edmisten and Brian Walton called a falconer who was a friend of the suspect to get the word out requesting the return of the stolen chicks. A returned call to Walton and Thelander led to the finding of the stolen birds in a pillowcase at the base of Morro Rock, near "Target Rock". The chicks were brought to the residence of the Davies late at night and then taken home by Walton to be cared for. The young falcons accepted offered food.

Young falcon chicks in Walton's living room

On the morning of May 10th, Captain Hugh Thomas from the California Department of Fish and Game (CDFG) organized a climb with Walton, Thelander, and Edmisten. Brian Walton returned the young to the nest. The adults fed the two young that day.

That night the young were stolen again by falconers from Southern California. The next day observers noticed the adults were not feeding the young, as before. Thelander climbed to the nest and found no young. Captain Thomas of CDFG surprised and arrested a climber with climbing gear and a walkie-talkie. Waiting in a Volkswagen below was a friend with a walkie-talkie. The two admitted climbing the Rock but the two young falcons were already gone. In September the two Bay-area residents were convicted in court of illegal climbing, but acquitted of the charge of attempting to steal the birds. This time the two falcon chicks, stolen by other falconers, were never recovered.

A 1972 article about the Morro Rock peregrine falcons published in the New Yorker magazine resulted in greater public awareness about the falcons and the yearly nest watch vigil.

The pesticide DDT was banned by the Environmental Protection Agency (EPA) from use in the United States in 1972; however, it was still widely used in some countries such as Mexico and in Central and South America. Among millions of migratory birds, including shorebirds and passerines returning northbound each spring from Latin America, were individuals carrying DDT pesticide loads that accumulated in their bodies from the foods they consumed. The pesticides also accumulated in the tissues of the peregrines that prey upon these migratory birds, subsequently resulting in eggshell thinning. Locally, persistent residual DDT levels present in the soils, stream deposits, and bodies of water were still a source of chemical contamination that "biologically magnifies" in concentration through the food chain. With the passage of time since the ban, decreasing DDT exposure was expected to have less eggshell-thinning impacts on 'top' predators like the endangered peregrine falcon, bald eagle, and brown pelican.

1973 The Davies continued to observe for the seventh year of volunteer watching. A pair of falcons was on Morro Rock in February, but one of them (the male?) apparently disappeared and no nesting was believed to have occurred. One day in mid-June, the Davies and John Schmitt observed four peregrine falcons at the Rock, two of which were clearly immatures. An aerial food exchange with one of the fledglings occurred. Later, Brian Walton and John Edmisten met Tom Cade, founder and director of The Peregrine Fund at Cornell University in New York. The three of them observed seven peregrines at Morro Rock that one day in June. Confusion exists about the origin of these birds: the young either fledged from Morro Rock after all or perhaps more likely from a nearby unknown eyrie.

The Fish and Game Commission declared Morro Rock an Ecological Reserve, making it illegal to climb above the ten foot level on the Rock without a permit from the California Department of Fish and Game. Signs were posted at the base of the Rock. Law enforcement is conducted by the California Department of Fish and Game, State Department of Parks and Recreation, and the Morro Bay Police Department.

The American Peregrine Falcon (*Falco peregrinus anatum*) was listed as an endangered species under the 1973 Federal Endangered Species Act. The peregrine falcon had also previously been listed on the State of California's first endangered species list.

1974 On May 2nd, the California Crime Technological Research Foundation, in cooperation with the California Department of Fish and Game, State Department of Parks and Recreation, and the Morro Bay Police Department installed electronic anti-trespassing devices on the Rock. The project was funded by the National Audubon Society and Defenders of Wildlife. Seismic sensing devices were placed on two access trails to detect footsteps. The devices sent radio signals to an antenna above police headquarters, with flashing lights and wailing honks to alert the dispatcher. Technical problems eventually ended the use of the devices; in particular, the impact of the waves against the Rock created a greater seismic disturbance than footsteps!

Four young falcons fledged from the nest.

Nest attendants: Ron Walker and John Glassburn, contracted by the CA Department of Fish and Game.

1975   Three eggs were laid; two young fledged at the Rock.

Nest attendants: Ron Walker, G. Adams

A Peregrine Symposium was held at the San Luis Obispo Police Department in September. Ron Walker and Ed Self led a symposium field trip to the Morro Rock Ecological Reserve.

Late in the year, the adult female was injured by a probable collision with a power line, causing a compound fracture of the wingtip (metacarpus bone). Morro Bay school children found the falcon flopping around on the ground, unable to fly, near a freeway on-ramp on December 11th. A California Department of Fish and Game warden obtained the bird and transferred it to the rehabilitation center at the Alexander Lindsay Museum, then to the Santa Cruz Veterinary Clinic.

1976   Morro Coast Audubon Society members presented engraved plaques, designed by Don Parham, to the students at a school assembly, in appreciation for recovery of the injured falcon found the previous December. The injured female falcon's wingtip circulation was impaired. Bone transplant surgery was performed on March 8th by Dr. James Roush, orthopedic surgeon at the Santa Cruz Veterinary Clinic. The last rib bone was used to bridge the break in the wingtip. The old falcon unfortunately could not be returned to the wild and eventually died in 1980.

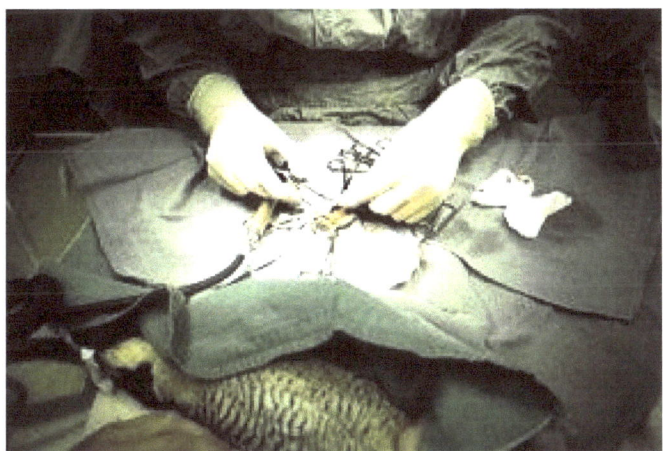

photo: SCPBRG files

Wingtip surgery

Meanwhile, the male had paired with a new mate. Three eggs were laid; one egg became addled (dead embryo within), two hatched and the young fledged.

Seven productive peregrine falcon eyries were known in the state.

An observation blind and a shelter for protection from strong, cold winds and foggy nights were constructed on the summit of Morro Rock.

Nest attendants: Merlyn Felton and John Glassburn, conducting a 24-hour surveillance.

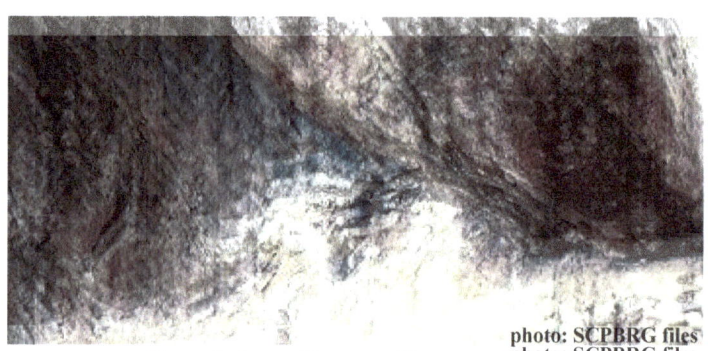

Female brooding small young, 1976 photo: SCPBRG files

1977 Nest attendant Merlyn Felton, under contract with CDFG, conducted 24-hour surveillance April 15th through July 5th. An improved observation blind and shelter combination was constructed. The observation blind across from the eyrie was entered at night to avoid disturbance and occupied to detect any attempted night raid on the nest. Nineteen persons were found illegally climbing the Rock during the season. Human disturbances were a continuing concern.

A live-trap was used to remove feral cats, which were delivered to the animal shelter. Felton's report discussed the potential threat of the increasing feral cat population at the Rock and made a recommendation for the cats' removal.

The eggs failed to hatch. Two 9-day-old prairie falcons were placed in the nest May 12th to maintain the parents' interest and nesting behaviors. The prairie falcons were removed May 21st. Two 11-day-old captive-bred peregrines were placed in the nest. They had hatched some 3,000 miles away in Ithaca, New York at The Peregrine Fund at Cornell University. The chicks had been flown by commercial airline to California, carried up the Rock in packs, and placed in the nest by Phyllis Dague of The Peregrine Fund. The adults accepted and fed the chicks, a successful "fostering" technique. The Morro Rock nest site was the first time and place in California where a wild eyrie was fostered with captive-bred peregrine falcon young.

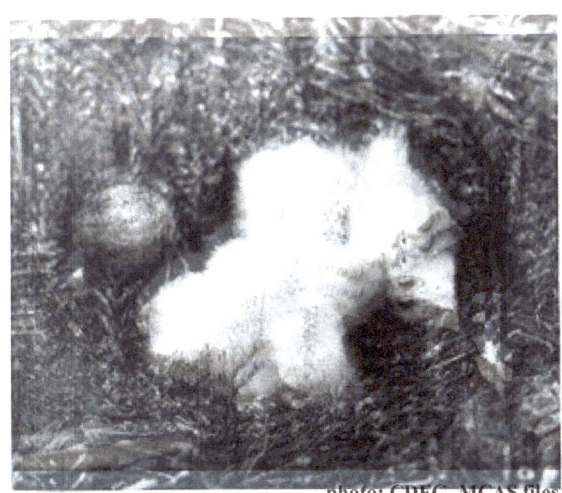

photo: CDFG, MCAS files
Prairie falcon chicks with peregrine egg to hold adults at the nest, 1977

photo: SCPBRG files
Phyllis Dague and Ron Walker place young peregrines into the nest, captive hatched at Cornell University, 1977

photo: CDFG, MCAS files
Banded fostered peregrine chick, 1977

The adult male was seen in the nesting area for the last time May 26th - he had appeared ill and weak the day before. The falcon had been wounded by a shotgun blast. It had been dead for several days when it was found on June 11th. X-rays revealed lead shot embedded in the wing.

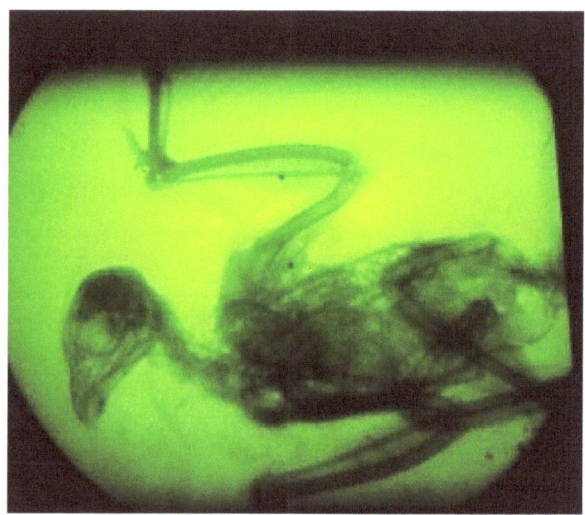

photo: SCPBRG files
X-ray of shot peregrine, 1977

The female had begun providing for the fostered young, capturing prey including two American kestrels and a hooded merganser, but mysteriously stopped hunting. A supplemental feeding program was started May 28th; however, the young male in the nest had died during the previous night, perhaps from exposure. Merlyn Felton imitated the male's food exchange vocalization to draw the female's attention. He then tossed live prey off the Rock - pigeons and young pheasants - for the female to capture and feed to the remaining nestling. The young fostered female fledged June 23rd. A new adult male peregrine showed up at Morro Rock July 5th.

Nest attendant: Merlyn Felton

Dr. James Roush, veterinarian and founder of the Santa Cruz Predatory Bird Research Group in 1975, spoke at the November meeting of Morro Coast Audubon Society about the peregrine falcon captive breeding program developing at UC Santa Cruz.

1978 The City Council on Feb. 13th adopted a Morro Coast Audubon Society proposal initiated by Don Parham, MCAS President, which designated the peregrine falcon as Morro Bay's City Bird.

Bob Mallette, California Department of Fish and Game biologist, spoke about peregrine falcon biology and conservation at a Morro Coast Audubon Society meeting.

None of the three eggs hatched; one disappeared, two were infertile. A red-tailed hawk chick was "cross-fostered" into the nest, died later, and was replaced with two prairie falcon chicks. An 18-day-old peregrine chick was placed in the nest May 31st - taken as an egg from a Monterey County nest and captive hatched in an incubator. Brian Walton climbed the Rock to place the chick in the nest, removing the older prairie falcons, and another prairie falcon chick of similar age was put in temporarily for company and warmth. Actor Mike Farrell ("B.J." on the MASH television series) was at the site to film a commercial for the California Department of Fish and Game non-game wildlife program. The fostered nestling later fledged successfully.

Nest attendant: Merlyn Felton

CDFG Captain Hugh Thomas removing infertile peregrine eggs

Brian Walton, Coordinator of SCPBRG, and Brian Farrell, Television Actor, filming CDFG Commercial

Construction of the captive breeding facilities continued at the UC Santa Cruz campus, within an old limestone rock quarry. Peregrine falcon breeding stock was obtained from Cornell University, licensed falconers, and from injured non-releasable falcons. Brian Walton became the Coordinator of this project called the Santa Cruz Predatory Bird Research Group (SCPBRG).

Captive breeding facilities under construction

Hyrum Strong, coordinator for the MCAS captive breeding fund, wrote an inspiring fund-raising article for the January 1978 "Flyway" newsletter. Contributions from many individuals and organizations have been an important source of non-government funding for the captive breeding project at UC Santa Cruz. $2,400 was donated to SCPBRG by the Morro Coast Audubon Society chapter.

Steve Schubert presented a peregrine falcon slide talk at the October meeting of Morro Coast Audubon Society. John Schmitt sold peregrine falcon lithographs with proceeds donated to SCPBRG. Brian Walton, coordinator of the SCPBRG project, was in attendance at the meeting.

A leg-banded juvenile peregrine (the surviving fostered 1977 female hatched at Cornell University) was found dead by three boys on November 5th, 1978 in the Bell Gardens river bed near Downey in Southern California, 200 miles southeast of Morro Bay. A local resident sent the numbered band and a few falcon feathers to the U.S. Fish and Wildlife Service Bird Banding Lab in Patuxent, Maryland. The falcon had been banded in the Morro Rock eyrie, before fledging, by Brian Walton and Dr. Eric Johnson, Ornithology Professor at Cal Poly, San Luis Obispo. Dr. Johnson later published an account of the Cal Poly band recovery in the journal "North American Bird Bander", since it was the first peregrine falcon to have been fostered into a nest in California and had survived for over a year.

1979 The falcons' nest was in an old cormorant nest constructed of seaweed. The first clutch of three eggs was removed March 21st. None of these eggs were successfully hatched in the incubator at Santa Cruz. Thin shells and dehydration made captive hatching difficult, although refinements made future hatching rates in the incubator more successful, where temperature and humidity could be carefully controlled. Captive incubation prevented possible breakage of eggs by the incubating adults, if the eggs had been left in the wild. The second clutch of three eggs was removed April 17th and kept in the incubator. One egg hatched. This was the first Morro Rock egg to be hatched successfully in captivity; also, this was the first year the site was "double clutched" as a management technique. Three infertile "dummy" prairie falcon eggs had been placed in the eyrie after the second clutch was removed, to hold the adults. The dummy eggs were removed on April 25th and two two-week-old foster peregrines were placed in the nest. They had been removed from a northern California eyrie as eggs and hatched in Brian Walton's home under 24-hour monitoring. Before being placed in the eyrie, the two chicks had been hand-fed ground quail with a peregrine puppet to prevent imprinting on human beings. The two fostered young later fledged successfully.

Nest attendant: Merlyn Felton

Twenty wild peregrine nesting pairs were known in California.

1980 Two clutches totaling seven eggs were laid. The first clutch of three thin-shelled eggs broke. The second clutch was removed (two eggs addled) and replaced with two artificial "dummy" eggs; one egg hatched in captivity. Later, two fostered chicks (captive hatched from wild eggs) fledged, appropriately on Independence Day.

Nest attendant: Merlyn Felton

The California Department of Fish and Game now contracted each year with the Santa Cruz Predatory Research Group for nest site surveillance.

Analysis of Morro Rock eggshells indicated one egg was 37% thinner than normal, the highest level of DDT-induced eggshell thinning ever recorded in a peregrine falcon!

1981 The falcons were again "double clutched", with a total of eight eggs laid in one season compared to the normal productivity of three to four eggs. Four eggs were addled; four hatched in captivity. Two fostered young (captive bred) fledged.

The nest guards' observation blind high up on the Rock was severely vandalized.

Nest attendants: Merlyn Felton, David Foote

Captive female feeding young to be fostered into the wild

1982   Seven eggs were produced by double clutching, the last year the Morro Rock nest site was manipulated this way (although in '80, '84, '91, '94, '98, '99 and '06 the falcons laid two clutches, without manipulation by SCPBRG). Two eggs were addled, five eggs hatched in captivity. One captive hatched nestling of Morro Rock origin was fostered at a reintroduction site on a ledge of a high-rise building in Westwood, but subsequently died after flying into a window. Two fostered young at Morro Rock (captive hatched from wild eggs) fledged.

Nest attendants: Merlyn Felton (7th year), John Schmitt

1982 eyrie 'The Altar'

Local author Harold Wieman wrote a pamphlet entitled "The Peregrine Falcons of Morro Rock", distributed by the Natural History Association of the San Luis Obispo Coast.

An opening ceremony event was held August 7th for the new peregrine falcon exhibit at the Morro Bay Museum of Natural History. This was a two-year, $15,000 project: sculptor- Dean Weldon; artist- Robert Reynolds; falcon specimens and eggs- Santa Cruz Predatory Bird Research Group; taxidermy- John Schmitt. The Natural History Association sold limited edition peregrine falcon prints by artist Robert Reynolds for fund-raising efforts. John McDonald, Morro Coast Audubon Society member, was chairman of the twelve-member NHA board. The MCAS chapter provided additional funding for the exhibit.

1983 The eyrie on the west side of Morro Rock was enhanced (improved structurally) with a layer of pea gravel placed by Merlyn Felton.

All four eggs were addled (dead). Three young were fostered (one captive bred, two hatched from wild eggs); one chick died in the nest. A helicopter flew very close to the Rock several times and may have caused the brooding female to be frightened from the nest, perhaps causing the nestling's death by exposure. Two young fledged.

Nest attendants: Eric Wise, Lee Aulman, Russell Thorstrom

1984 Breakwater repair near the Rock was postponed in 1983 and 1984 until after the breeding season, following damages to the rock jetty by El Nino winter storm surges.

Most of first clutch was broken. Two addled eggs were collected after abandonment of the nest. A second clutch was laid at a relocated eyrie for a season total of six eggs produced; two eggs hatched in captivity. Two fostered chicks, one male and one female, were placed in the nest, but were not genetic siblings (see 1987 information). The fostered female nestling originated from captive-bred parents at the Santa Cruz facility and the male as a wild egg taken from a nest along the Big Sur coastline of Monterey County and hatched in captivity.

Nest attendant: Janet Linthicum

1985 None of the three eggs removed hatched in captivity. The dummy eggs were removed April 29th and replaced with one seven-day-old peregrine chick. This chick was removed May 12th, fostered into another eyrie along the coast in San Luis Obispo County, and replaced at the Rock -appropriately on Mother's Day!- with two ten-day- old fostered female chicks (captive hatched from wild eggs). Upon fledging from the eyrie June 11th one juvenile fell to nearly the base of the Rock, was harassed by the gulls, and was not fed by the parents. The fallen fledgling accepted a whole blackbird provided by the attendant June 13th and was captured June 14th by Thompson and Wise, then carried up the Rock in a backpack and released. It was later fed by the parents. Two young fledged.

Nest attendant: Dean Thompson (and his dog Desmond!), with occasional assistance from Eric Wise, John Schmitt, Jamey Eddy, Gary Guliasi, Staci Kawa, and Morley Weir. The location of the eyrie on the south face above the parking area made it possible for curious visitors to view the falcons through a spotting scope and ask questions about the falcons, expanding the nest attendant's role by providing information to the public about peregrine falcon biology.

Fostered youngsters in 1985 'diving board' eyrie

1986 Three eggs were laid; one hatched in captivity. Two fostered young (captive bred) fledged.

Nest attendant: Andre Wille

1987 The two genetically unrelated fostered peregrines of 1984 became this year's new breeding pair on Morro Rock (this adult female was perhaps still present through the 1991 nesting season). Four eggs were laid; three were broken in the nest, one dead on arrival at Santa Cruz. Two fostered young (captive bred) fledged.

Nest attendant: Matt Nixon.

1988 Five eggs were laid; two hatched in captivity. Two fostered young (captive bred) fledged.

The adult male sheared off a wing due to a wire collision. He was found near Toro Creek to the north of the Rock and delivered to the Central Coast Rehabilitation Center, but did not survive.

Nest attendant: Becky Pierce.

1989 Three eggs were laid; neither of the two eggs removed hatched in captivity. Two fostered young (one captive bred, one captive hatched from a wild egg) fledged.

Nest attendant: Meighan O'Brien

1990 Four eggs were laid; one hatched in captivity. One fostered young (captive bred) fledged.

Nest attendant: Kit Crump.

There were 106 known active sites (breeding pairs of peregrines) in California. One hundred wild young fledged with no human intervention at all. Eighty-one young peregrines were fostered, cross-fostered with prairie falcons, or released from hack boxes by the Santa Cruz Predatory Bird Research Group.

1991 The first clutch broke with an unknown number of eggs. A single egg (from continuation of laying of the first clutch or perhaps from a second clutch?) was removed at about 14-days old on April 12th; dead on arrival at SCPBRG. Three dummy eggs were placed in the nest; later replaced on April 26th with two fostered young (one captive bred, one captive hatched from a wild egg). The two fostered young fledged the last week of May.

Nest attendant: Christy Craig

Nest site attendant Christy Craig

The April issue of National Geographic magazine featured an article written by Galen Rowell entitled "Falcon Rescue", including a graph depicting high toxin residues measured within the eggs of the Morro Rock falcons.

1992 The breeding female falcon at the Rock was trapped by Lee Aulman and Scott Francis and identified by its SCPBRG leg band: taken as a wild egg from a nest in northern California, captive hatched, and fostered into a Big Sur coast nest in 1984. This confirmed that the 1992 breeding female was a different individual and had replaced the banded female identified in 1987. A second leg band was placed on the trapped female before release (SCPBRG blue band on the left leg, black band on the right leg). Note: a wild egg was taken from the same Big Sur eyrie in 1984 and captive hatched. This chick was then fostered into the Morro Rock eyrie, and later became the 1987 breeding male!

The eyrie was located on the south face of the Rock, facing the harbor entrance. This nesting pothole, with a "diving board" rock perch at the entrance, was previously occupied by breeding peregrines in the years 1985 and 1971. Four eggs were laid; all dead on arrival at Santa Cruz. Two young were fostered April 22nd and later fledged. One fostered chick originated as a wild egg from the Channel Islands and was captive hatched at Santa Cruz, the other was hatched from captive bred falcons at The Peregrine Fund facility in Boise, Idaho.

photo: Steve Schubert
Agitated peregrine in flight as climbers foster young into the 'dving board' eyrie, 1992

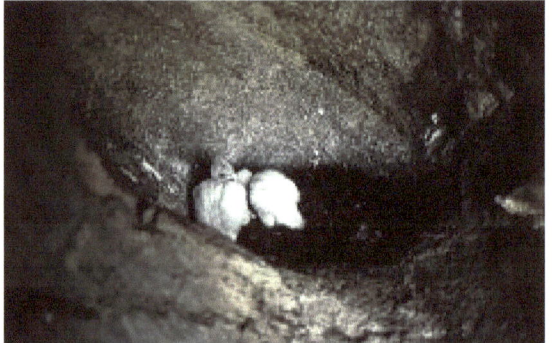

photo: Brian Latta, SCPBRG files
Peregrine chicks fostered at Morro Rock, 1992

Nest attendants: Christy Craig, early in the season, followed by an organized daily watch of Morro Coast Audubon Society volunteers: Eilleen and Charles Bowen, Barbara and Ernie Eddy, Joanna Frawley, Rich Hansen, Connie and Marlin Harms, Jim and Karen Havlena, John McDonald, Bill and Emma Moore, Don and Jo Parham, Steve Schubert, Lisa Trayser, Morley and Douglas Weir

The watch schedule was organized by Frank Little, MCAS President.

A two-part article about the Morro Rock falcons, written by Steve Schubert, was printed in the local newspapers in April for the "State of the Bay" series.

1993 The Santa Cruz Predatory Bird Research Group program halted many of its management activities after last year's breeding season. A "stop, wait, and see what happens" approach was deemed necessary to determine if the peregrine falcon population was self-sustaining and increasing on its own. The management program in California - captive breeding, double clutching, captive hatching wild eggs in incubators, fostering and cross-fostering young into wild nests, and reintroduction of peregrine falcons by the hacking technique - successfully increased the number of fledglings in the wild by over 750 birds. Monitoring known active nest sites and surveying for new occupied breeding territories would be continued. Concern was raised that peregrines at some sites might still suffer from reproductive failures as these reintroduced birds became adult breeders.

The Morro Rock falcon pair was present in late February and March, occupying the pothole caves again on the south face, engaging in courtship food exchanges and mating. Nest site selection, egg laying, and incubation exchanges were taking place by late March.

By early May the tiercel brought prey to the eyrie while the female was inside the cave and vocalizing, indicating there were wild-hatched young in the nest being fed. The nest was not manipulated with 'hands-on management' for the first time since the Morro Rock peregrines had last successfully hatched and reared their own young, in 1976!

The food begging vocalizations of the eyasses heard from within the cave May 18th confirmed their presence. A nestling's small head peered out from the edge one week later. Downy feathers on the head and developing black malar stripes below the eyes indicated the age at about three weeks old. During late May through early June the tiercel dropped prey into the eyrie for the chicks to feed on without assistance.

Two young fledged June 12th. The fledglings developed their flying skills, received prey in aerial food exchanges from the parents until becoming more independent, and dispersed from the natal area during the late summer.

After the young fledged, eggshell fragments and feathers were collected from the nest site. The feathers of prey remains were identified by John Schmitt as one individual of a warbler species, two Western sandpipers, one European starling, one Rock dove, one Cliff swallow, and probable windblown feathers of a Pelagic cormorant, Brown pelican, and Western gull. The eggshell analysis conducted by the Western Foundation of Vertebrate Zoology determined there was 23.6% thinning.

Occasional volunteer nest site observers: Judy Sullivan, Pete Lambert, Steve Schubert, Paula Becker, Jamey Eddy, John McDonald, Morley Weir

1994 A peregrine falcon slide talk was presented by Steve Schubert at the April meeting of the Friends of the Estuary organization, in the town of Morro Bay. Merlyn Felton discussed his experiences as a nest guard for many years and his in-depth knowledge and observations of the Morro Rock falcons. Felton brought a captive peregrine falcon named Glass to the meeting, to the awe of more than 150 persons in attendance.

The falcon pair occupied the large potholes high on the south side of the Rock, above the parking area. This was the third consecutive year that nesting occurred on this steep south-facing escarpment. During the 1970's and 80's the nest sites were most frequently located on the west face of Morro Rock high above the sea, although nest sites were also located on the south face in 1972 (when the young were stolen) and in 1985, and on the east face in the late 1960's. The falcons repeatedly utilized a number of favorite perch sites, feeding stations, and food-caching potholes throughout the season. With the use of a spotting scope, both adults were observed to be leg-banded. The female's blue band (left leg) and black band (right leg) revealed it was the same breeding falcon trapped on the Rock in 1992. The male was also banded with a SCPBRG blue band (right leg).

Incubation of eggs apparently began during the last week in March, but the first clutch was lost and the eyrie was abandoned. The pair "recycled" during a several week period, renewing courtship and mating activities. They selected a different nest site in a small pothole below the '92 'diving board' eyrie. Incubation of the second clutch began by the last week of April. On May 30th the male exchanged a shorebird with the female, which was taken into the eyrie. Small plucked feathers were soon drifting out on the breeze. This was an indication that hatching had successfully occurred - food was being brought to the eyrie and fed to the young. Unfortunately, the one known downy chick disappeared at three weeks of age and may have been lost to predation or perhaps fell from the edge of the eyrie.

Brian Latta and Jamey Eddy climbed to the empty eyrie July 1st and collected small eggshell fragments for laboratory analysis of eggshell thickness.

Occasional volunteer nest site observers: Judy Sullivan, Pete Lambert, Steve Schubert, Jamey Eddy, John McDonald, Morley Weir (a yearly observer since 1985)

photo: Steve Schubert
Adult peregrine at pothole eyrie edge, 1994

Following concerns expressed to State Park personnel about the feral cat population thriving at Morro Rock, signs were posted with reference to the penal code violations for abandonment of animals and feeding of wildlife. The Morro Bay City Council approved the contract agreement to spay and inoculate feral city cats, including those at the Rock. Volunteers, working with State Parks Resource Ecologist Vince Cicero, live-trapped many of the feral cats at the Rock, including diseased animals and abandoned kittens, many of which were later adopted. Without continued live-trapping efforts and enforcement of violations, however, the cat population may increase again and threaten the native wildlife due to predation.

The book entitled Falcons of the Rock was published, authored by pen name Donovan Lavender. This publication provides a detailed and personal account of observations of the Morro Rock peregrine falcons over a several year period.

In December, live trapping of leg-banded adult falcons along the coast resulted in some interesting findings. The breeding male at an eyrie at Point Arguello on Vandenberg Air Force Base was identified as having been captive-bred and fostered into the Morro Rock nest in 1991. The breeding female at Point Arguello had originated as a wild fledgling from the Channel Islands. This pair had lost two eggs the previous breeding season and were not successful. At the Piedras Blancas eyrie north of San Simeon, the breeding male had been captive-bred and fostered into the Morro Rock nest in 1990. The Piedras Blancas breeding female had originated as a wild fledgling from the interior of northern California near Ukiah. This pair had successfully fledged two young the previous nesting season. The dispersal of Morro Rock's fledged peregrine falcons north and south along the coast and their eventual establishment as breeders at other nesting territories are indications of the success of the captive breeding and reintroduction program, its goals to supplement and increase the imperiled peregrine falcon breeding population which had suffered more than a 90% decline in California during previous decades.

1995 The adult male peregrine from the Piedras Blancas eyrie was found injured on the grounds of an elementary school in Cambria on February 27th. This bird had been fostered into the Morro Rock eyrie in 1990. After being transferred to a local veterinarian and then to personnel at the SCPBRG project, it was deemed necessary to euthanize the falcon, which had five breaks in its wing bones probably caused by a flight collision with a power pole wire in the town.

In early April a request was made to the U.S. Coast Guard to temporarily halt repairs to the foghorn structure at the end of the Morro Rock breakwater, as the use of a helicopter to airlift building materials was involved. The Coast Guard and its private contractor agreed to resume repairs without the use of a helicopter, to eliminate this potential disturbance so early in the falcons' nesting season.

The continued live-trapping of feral cats at Morro Rock during a one-year period resulted in the removal of at least forty-five cats!

The same pair of falcons returned to Morro Rock to breed in 1995 (identified by their leg bands). The nest was located on the seaward (southwest) face of the Rock, at or very near the same eyrie occupied in 1991. Observations of the falcons' nesting activities were made from the breakwater jetty. Hatching of eggs occurred by early May, and two young- a male and a female - fledged successfully from the nest by mid-June. Eggshell fragments and the feathers of prey remains were collected from the eyrie. John Schmitt identified the prey feathers as those of a Bonaparte's gull (1), European starlings (2), Killdeer (1), House finch (1), Marbled godwit (1), Mourning dove (1), and Cliff swallow (1).

Occasional volunteer nest site observers: Judy Sullivan and Steve Schubert.

photo: SCPBRG files
One of many chicks fostered at Morro Rock

photo: SCPBRG files
Falcon's eye view

1996 After nesting on the seaward face of Morro Rock the previous year, the nest site was relocated by the same falcon pair to the south-facing side, facing the harbor entrance inside the breakwater. The eyrie was in the same oblong-shaped pothole as that used in 1994. Eggs were laid by the last week of March, as evidenced by the female and male exchanging incubation duties. One male nestling successfully fledged in early June.

Mitch Siemens collected eggshell fragments from the eyrie in September.

Occasional volunteer nest site observers: Judy Sullivan, Steve Schubert, Gloria Rasmussen

1997 An interpretive signboard was installed at the base of the Rock near the south side parking area, with information about the geology of Morro Rock and a description with several photographs about the resident peregrine falcons. Development of this educational "kiosk" was coordinated by the efforts of Harold Wieman, local author and volunteer with the Central Coast Natural History Association and a long-time observer of the falcons at the Rock.

The same falcon pair nested in the same pothole that was used in 1994 and 1996. One nestling hatched in late April. At about 35-days-old on May 28th the nestling fell and scrambled down the rock face, gripping and coming to perch in a small pothole some 60 feet below the eyrie. After seeking shelter and scrambling about on the cliff face below the eyrie for several days, wailing at the attending parents flying and perched nearby, the fledgling finally took a short flight of a few feet on June 1st. Several strong flights along the face of the Rock were made during the next several days and on June 5th the fledgling was already pursuing swallows.

Occasional volunteer nest site observers: Judy Sullivan, Steve Schubert, John Edmiston

Within San Luis Obispo County there were at least six coastal and one inland peregrine falcon nesting site (within the Santa Lucia Wilderness Area of Los Padres National Forest), although several of these falcon pairs did not successfully raise young this year. It was estimated that more than 150 nesting territories throughout the state were occupied by peregrine falcons, a dramatic increase over the past several decades.

1998 Tom Cade, founder of The Peregrine Fund at Cornell University in 1970, visited at the Rock one day during the spring and expressed his appreciation to Judy Sullivan for her monitoring efforts. Judy was observing the falcons that day as a volunteer, a yearly commitment and passionate pastime.

During the nesting season, KSBY-TV reported on the Morro Rock peregrine falcons for a television news story.

After several members of Morro Coast Audubon Society and the local public expressed concerns about the proposed launch site for the Fourth of July fireworks, the roundtable planning committee agreed to relocate the launch site a greater distance from Morro Rock, from a floating barge further back in the bay; nevertheless, during the nighttime fireworks display a volunteer MCAS observer noted the crowded vehicle parking and illegal fireworks set off by individuals at the parking lot near the base of the Rock. The potential disturbances and unknown stressful impacts on the falcons and other wildlife should be taken into account by the fireworks planning committee and monitored from year to year, including patrolling at the Rock by law enforcement personnel during the fireworks event.

The falcon pair, presumably the same adult male and female identified since the 1992 nesting season, failed to hatch eggs in two different eyries: the first clutch laid in the oblong-shaped pothole occupied previously in 1994, 1996, and 1997; and the second failed nesting attempt in the pothole with the 'diving board' perch, last occupied in 1993 (note: a 1972 photograph of peregrine nestlings within this same 'diving board' eyrie is on file with SCPBRG).

An ascent of Morro Rock in August was made by Brian Latta, Lee Aulman, and Steve Schubert. Latta rappelled by rope down to the two eyries and removed one egg from each nest. The egg from the lower eyrie was partially broken and empty and the second intact addled egg from the higher 'diving board' eyrie was perhaps infertile and never hatched. A KSBY-TV reporter and a cameraman covered the event for the evening news broadcast.

The average eggshell thinning for the collected Morro Rock eggs, measured by the Western Foundation of Vertebrate Zoology, was 17.6%. Average thinning of eggshells collected at Morro Rock from 1976 through 1998 was 21.4%.

Occasional volunteer nest site observers: Judy Sullivan, Steve Schubert

1999 The resident female peregrine falcon along the coastline at Bixby Creek was trapped by Brian Latta and identified by its leg bands. This falcon had been fostered as a nestling into the Morro Rock eyrie in 1983 (hatched from captive bred parents at SCPBRG), and as an elderly 16 year-old adult now nested along the rugged Big Sur coast approximately 100 miles to the north of Morro Bay.

The resident Morro Rock peregrines began nesting in mid-March in the eyrie with the "diving board" perch, also used the previous year, but failed to hatch clutches of eggs in apparently two different nesting attempts. Three falcon chicks were available for fostering into the eyrie at this time - this was the first time since 1992 that chicks had been fostered at the Morro Rock nest site. A private falconer provided the chicks, hatched from captive bred parents that were originally SCPBRG breeding stock. On April 22nd the three 25-day old peregrine chicks were leg-banded and placed into carrier boxes.

The ascent to the summit of the Rock was made by Brian Latta and Craig Himmelwright. Elizabeth Hoyt, Martine Lynch, and Steve Schubert assisted in the climb by bringing up the climbing gear and the carrier boxes. Latta and Himmelwright rappelled by rope down the cliff face and deposited the chicks into the eyrie.

All three nestlings - one male and two females - fledged from the nest on May 6th and rapidly developed flying skills during the days that followed.

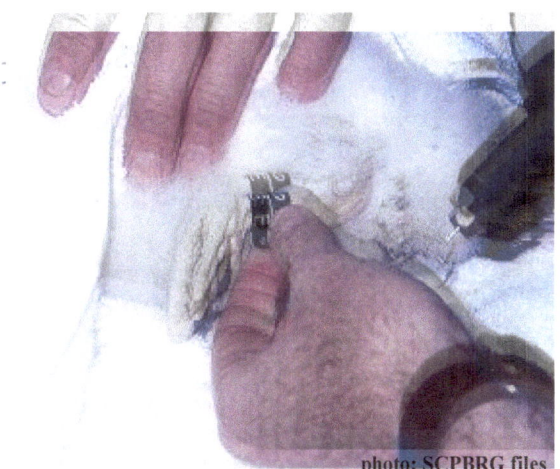

Latta leg-banding fostered peregrine chick

Chicks in carrying box

Himmelwright rappelling to eyrie to foster 3 chicks

Agitated adult peregrine

On May 21st the adult female peregrine was found dead on the ground at the base of Morro Rock - she had survived in the wild to an old age of 15 years and had been the resident breeding female at Morro Rock since at least 1992. The falcon was taken to SCPBRG for eventual transfer to a lab for necropsy. Out of concern for the fledgling falcons' ability to survive due to the loss of a parent, a supplemental feeding program began on May 23rd. During the next several weeks, licensed falconer John Edmisten released domestic pigeons at the Rock. The larger of the two fledgling females became adept at catching released pigeons, and her sibling would also feed with her. The young male appeared to be bringing in his own prey successfully. The young falcons eventually dispersed from their natal site at the Rock.

Brian Latta's written account of fostering the falcon babies at the Morro Rock nest on Earth Day was posted online at SCPBRG Field Notes:

http://www2.ucsc.edu/-scpbrg/field.htm

Occasional volunteer nest site observers: Judy Sullivan, Steve Schubert

On August 20th a delisting ceremony was held at the World Center For Birds Of Prey in Boise, Idaho. The American Peregrine Falcon was removed from the federal endangered species list. In attendance were the Secretary of the Interior, staff and volunteers of The Peregrine Fund, and members of the media providing television and news coverage. The delisting was heralded as a successful example of the 1973 Endangered Species Act at work. At the time the peregrine falcon continued to be designated by the California Department of Fish and Game as a state listed endangered species and fully protected by federal and state laws.

2000 A new pair of peregrine falcons became residents at Morro Rock this year. The old female - 'Morro Mary' - died in 1999, and the male later disappeared also. On May 14th, Brian Latta from SCPBRG used a high-powered spotting scope to observe the leg band numbers on the two new resident falcons. They were identified by their bands.

The new male named 'Rudy', two years old, was hatched in a wild nest along the coast at Point Arguello on Vandenberg Air Force Base in Santa Barbara County. The father of this new male was captive bred and fostered into the Morro Rock nest in 1991, and is now a member of the breeding pair at Vandenberg. The mother of the new male was banded in 1992 as a wild chick in a nest on the Channel Islands, and his maternal grandmother was captive bred and fostered into a Big Sur nest in 1986.

The new female named 'Milli' (in honor of the new millennium and also the Star Wars movie ship the Millennium Falcon), two years old, was released in 1998 at the Sudden Flats hack site, also located at Vandenberg Air Force Base and south of the active wild nest site. According to Janet Linthicum, SCPBRG, "she was removed as a wild nestling from the Cal Fed Building in Los Angeles because of dangerous fledgling conditions there". The new female falcon's parents were unbanded and unidentified.

The pair began egg-laying in March and successfully hatched young during the last week of April. Interestingly, they nested in the same pothole with the protruding "diving board" used by the previous older pair. This nest site has been documented to have been occupied by nesting peregrines intermittently back to the early 1970's.

On May 19th, Brian Latta, SCPBRG, banded two wild chicks about 25 days old - both males - with black and silver leg-bands. Marlin Harms, Steve Schubert, and Joan Rainey-Day assisted in carrying ropes and climbing gear to the rappelling location on the Rock. Latta collected a crushed egg from the eyrie as well as prey remains for later identification. Several hundred feet below at the base of the Rock many observers watched with binoculars, spotting scopes, and telephoto camera lenses, some reclining comfortably in lounge chairs!

On May 31st, the two peregrines successfully fledged from the nest.

A pair of Great horned owls nested and raised young for the second consecutive year on a rocky talus slope and cliff face on the east side of Morro Rock. The falcons and owls are apparently coexisting in relative close proximity, although Great horned owls are potential predators of young peregrine falcons.

Occasional volunteer nest site observers: Judy Sullivan, Steve Schubert, Vince Cicero, Leslie Thompson

2001 Early in the year peregrines were observed perching in a pothole on the north side of Morro Rock, and during the nesting season falcon watchers became suspicious that there was more than one pair of falcons residing at the Rock. On May 7th, Brian Latta, Janet Linthicum, and Brian Walton from SCPBRG made the long drive from Santa Cruz and made observations confirming there were two pairs of peregrine falcons, residing on the south and north sides of Morro Rock! Subsequently, observers started communicating with hand-held radios, making simultaneous observations of the falcons on both sides of the Rock. This is the first known occurrence, since the yearly nest watch began in the late 1960's, of more than one pair of peregrine falcons nesting on Morro Rock (note: in 1970, only two known peregrine falcon nest sites were found in the entire state of California!).

The south side pair apparently attempted to nest, but was unsuccessful and no young were produced this year. The north side pair, nesting in the pothole, hatched 3 chicks. On May 14th the chicks - one male and two females - were banded in the nest by Brian Latta, who rappelled by rope from a height above the eyrie. Assistants carrying climbing gear and also making the climb were Vince Cicero and Mike Walgren from State Parks, Steve Schubert, and Mike Baird. Latta removed prey remains from the eyrie, and among the feathers and bones placed in a bag for later identification was an unplucked adult male Wilson's warbler that had not been fed upon.

That day a partial read on the north side adult female's leg-band numbers was made using a high powered spotting scope (she had been named 'Xena', and eventually in 2004 all of her leg-band numbers were read successfully).

The three north side chicks successfully fledged from the nest May 29th.

North side banded female peregrine in flight

The south side pair this year were observed a number of times hunting in tandem, often several times a day, attacking and feeding on gulls at the end of the sandspit. The frequency of attacks on gulls was surprising, compared to previous years monitoring the falcon's hunting behaviors. In retrospect, this was a new adult female which had paired with the male "Rudy", and the two attacked gulls in tandem more frequently than the pair the year before. It was eventually noticed this new resident female on the south side was lacking leg-bands, confirming the change in the pair bond since the prior year ('Milli', residing on the south side the year before, was leg-banded). Without these leg-banding observations, it would not be possible to always conclusively identify the falcons from year to year, due to occasional losses and replacements of mates from time to time. These mate replacements are speculated to be from the 'floater' population of other peregrines in the area.

During the summer of 2001, following the nesting season, once again a new female peregrine took up residence with the male 'Rudy' on the south side of Morro Rock, apparently replacing the adult female that had recently disappeared. The newly arrived young peregrine finished molting during the summer into adult plumage. This new resident female was unbanded.

Occasional nest site observers: Judy Sullivan, Steve Schubert, Bob Isenberg, Roy and Ella Burke, Vince Cicero and Leslie Thompson, Curt and Sandra Beebe, Mike and Heidi Baird. A gathering in honor of Judy Sullivan - in appreciation for her many years of efforts monitoring the Morro Rock falcons - was attended by a large group of friends and falcon watchers. Judy moved to Seattle (but would move back more than 2 years later and resume her daily morning vigils watching the falcons at the Rock).

2002 Two pairs of peregrine falcons nested and successfully fledged young at Morro Rock! There are few documented places on the North American continent where territorial peregrine falcons are known to nest in such close proximity.

The north side pair - named 'Xena' and 'Zephyr' - nested in a less conspicuous pothole with a grassy ledge located to the left (east) and above the 2001 pothole eyrie. The south side pair nested in the 'diving board' eyrie.

Three chicks fledged from the north side eyrie the third week of May. Several days later one chick fledged from the south side eyrie.

Occasional volunteer nest site observers: Bob Isenberg, Steve Schubert, Roy and Ella Burke

2003 During the winter, both of the previous year resident males - 'Rudy' and 'Zephyr' - had disappeared form unknown causes. By the nesting season the lone south side female had pair-bonded with a new mate. There were clearly three, not four, peregrines now residing at Morro Rock. Well into the nesting season the new male mated with and brought food to both females, on both sides of the Rock. There was speculation that the new male might attempt to simultaneously nest at both locations and rear two broods of chicks. Eventually, he nested and reared young on the south side only. The east side female did not nest this season, although the male falcon would occasionally perch close by to her, after bringing food to the other adult female and chicks in the eyrie on the south side.

The south side pair successfully fledged two young - one male and one female - that hatched in early May and fledged from the 'diving board' eyrie on or near June 10th.

Visitors to the Rock this season included a 36-year-old Los Osos resident with his children. He recalled that when he was about eight years old he was with the group of Morro Bay elementary school children who had found the injured female peregrine falcon on the ground near Hwy. 1, in the year 1975, and they rescued the falcon (see the text for the years 1975 and 1976). Retired Fish and Game Biologist Bob Mallette was also visiting one day, having traveled with friends from Sacramento, and was enjoying watching the falcons at the Rock. He had spoken about the peregrine falcon recovery efforts as a guest speaker at a Morro Coast Audubon Society chapter meeting in 1978.

Occasional nest site observers: Bob Isenberg, Roy and Ella Burke, Steve Schubert, Gary Robertshaw, Jeff Sipple, Jerry Franklin, Tom Roff, Diane Schoditsch

The U.S. Fish and Wildlife Service initiated a post-delisting survey of the peregrine falcon population. In California, a random sample of nest sites to be monitored was chosen from peregrines eyries known to have been active within the last 5 years. Unexpectedly, both Morro Rock nest sites were randomly selected for the nesting survey. Volunteers will play a large role monitoring nest sites and submitting data sheets (with information on nest site occupancy and nesting success), conducted every three years through 2015.

photo: Gary Robertshaw
South side adult falcon pair

2004 In February, a Valentine's Day celebratory falcon watch was attended by a large group of local volunteers and two staff members from SCPBRG. Two pairs of peregrines were observed that day at the Rock. Only one of these four falcons was leg-banded.

Brian Latta used his Celestron scope and clearly read the ID numbers on the leg-band of the east side female. A cellular phone call was made to a staff person at the SCPBRG office at Long's Marine Lab in Santa Cruz, reporting the band numbers so they could be looked up in the office files.

This adult female 'Xena' had nested successfully in 2001 and again in 2002. She did not nest in 2003, but associated with the south side male much of the nesting season. She then pair-bonded with a new male by the time of the 2004 nesting season. This year the pair perched more frequently and then nested on the east side of the Rock. The south side adults were now named 'Khaos' and 'Elvis', the adults on the east side were 'Xena' and 'Esteban'.

photo: Gary Robertshaw
Fledgling peregrines

Steve Schubert observing the falcons at the Rock

Both pairs successfully fledged young. The east side pair fledged two young - one male and one female - in mid-May. About one week later the south side pair fledged three young from the 'diving board' eyrie.

Occasional nest watch observers: Judy Sullivan, Bob Isenberg, Steve Schubert, Roy and Ella Burke, Vince Cicero and Leslie Thompson, Cleve Nash, Gary Robertshaw, Vince and Rosemary Duffy, Jeff Sipple

2005 Two pairs of peregrine falcons resided at Morro Rock again this year. Three nestlings - one male and two females - fledged from the south side 'diving board' eyrie on or about May 16th. The east side pair attempted earlier in the season to nest on a ledge to the right (north) of the 2004 eyrie, but failed to produce young this year.

Occasional nest watch observers: Judy Sullivan, Bob Isenberg, Steve Schubert, Roy and Ella Burke, Vince Cicero and Leslie Thompson, Cleve Nash, Gary Robertshaw, Vince and Rosemary Duffy, Jeff Sipple

State Parks granted permission for a Native American group to climb Morro Rock on the summer solstice, at a time when peregrine fledglings are present and at the height of the gull nesting season. Permission was also given for the lighting of a ceremonial bonfire at night high up on the Rock. Input from local falcon observers and staff from SCPBRG was given at a discussion meeting with State Parks, with concerns expressed about potential disturbances to wildlife and the precedent of granting approval to climb at a protected Ecological Reserve, where climbing restrictions have been in place since the 1960's. Careful monitoring of these future climbing activities has been requested.

photo: Cleve Nash
Nestlings at the 'diving board' eyrie

Food delivery to nestlings at the 'diving board' eyrie

2006 The 10th annual Morro Bay Winter Bird Festival, held during the month of January, continued each year to offer a peregrine falcon slide talk and field trip to Morro Rock.

Two pairs of peregrine falcons nested on Morro Rock. The south side pair nested in the pothole eyrie - located below the 'diving board' eyrie - last occupied in 1998. Eggs were laid during the last week of February, relatively early in the nesting season compared to previous years of observations. The other pair of falcons nested on the north side at the same eyrie last occupied in 2002, located on a grassy ledge. Both pairs were incubating eggs by the first week in March, but failed in their first nesting attempts. The north side pair had taken food into the eyrie apparently to feed one or more young that had hatched on their first nesting attempt, but feeding by the adults eventually stopped and the eyrie was abandoned after loss of the young. The north side pair relocated to a more east facing eyrie located on a wide ledge, but failed again the second nesting attempt. Meanwhile, the south side pair 'recycled', relocating to the 'diving board' eyrie in their second nesting attempt and successfully fledgling one young on the late date of July 4th, Independence Day.

Occasional nest watch observers: Judy Sullivan, Bob Isenberg, Steve Schubert, Gary Robertshaw, Cleve Nash

photo: Cleve Nash
Peregrine fledgling

photo: Cleve Nash
Fledgling taking flight

The second in a series of U.S. Fish and Wildlife Service post-delisting peregrine falcon nesting surveys was conducted in 2006 (taking place every three years until the year 2015). Monitoring forms were completed and submitted to the USFWS for nesting observations at the two Morro Rock eyries.

In addition to assisting USFWS with the survey, Santa Cruz Predatory Bird Research Group staff and volunteers conducted a state-wide census of nesting peregrines, surveying more than 200 known historic peregrine nesting sites to obtain breeding pair occupancy data, an effort similar to the nesting surveys of the 1970's. A summary of this survey, posted on the SCPBRG website, follows:

2006 Peregrine Falcon Survey Results
Since the peregrine falcon was removed from the Federal Endangered Species List, the US Fish and Wildlife Service has been conducting a post-delisting monitoring survey every three years of a sample of known territories across the country, to assess whether delisting is having a negative effect on peregrines. SCPBRG coordinates the effort for California, monitoring 34 survey eyries. We ask individuals, both private and agency, including SCPBRG personnel, to monitor survey sites in their area to a standard protocol, and report results to us. Thirty-four individuals provided data for the survey this season, which was then shared with USFWS.
We took the opportunity of 2006 being a survey year to also attempt to visit as many other known and suspected peregrine territories in California as possible. No statewide survey has taken place since 1992, when the last multi-agency survey documented 113 pairs in the state. We picked a difficult year for this effort, not planning on record amounts of rain and snow into the late spring, and very high fuel prices. A grant from Newman's Own helped us fund the effort. Over 60 public and private individuals participated in the survey, many at their own expense of time and travel. We appreciate the effort of each one of them.
Some sites were not accessible until July due to snow on the ground. To date, 271 sites in California have been identified as having an active breeding pair at least once since the recovery began in the 1970s. 236 known or suspected sites were visited this year, and 215 provided some useful data. 167 sites had at least one adult present, and 154 had a confirmed active pair. Nearly 30 sites were newly discovered, or reported to us for the first time by individuals who have known of them for some time. The average fledge rate at successful nests was approximately 2 young per pair, with a minimum total of 146 young produced. Since we know many territories were not visited during the nestling stage, far more young peregrines fledged in the state this year. Our focus was not productivity, but a census of pairs.
Unusual eyries this season included two in abandoned stick nests on transmission towers in the San Francisco Bay Area, one on a small dike in a salt works, and two in old-growth snags above a forest. With the absence of widespread decimating contamination, and remarkable adaptability, it's becoming difficult to know where to look for peregrines!
Again we thank every individual who participated in this year's survey efforts: Allison, Brian; Anderson, Steve; Andreano, Paul; Aulman, Lee; Baird, Mike; Ball, Cathy; Bell, Doug; Bennet, Joe; Boyd, John; Cottrell, Kanit; Davis, Jeff ; Derby, Debbie; Dexter, Ken; Duffy, Vince; Dunlop, Nick; Eakle, Wade; Estes, Don; Francis, Scott; Garrison, Dennis; Gist, Jessica; Grant, Grant; Gregoire, David; Guliasi, Gary; Hamm, Kieth; Haschak, Art; Himmelwright, Craig; Holm, Greg; Hunt, Terry; Keiffer, Bob; Kirven, Monte; Latta, Brian; Linthicum, Janet; Lish, Cheryl; Loewen, Evet; Martin, Ryan; Maurer, Jeff; Menzel, Sandra; Morse, Neil; Nelson, Laura; Neville, Glenn; Oliveri, Joe; Pagel, Jeep; Rich, Adam; Robertson, Mark; Rowlette, Richard; Schubert, Steve; Sipple, Jeff; Smith, Zack; Sooter, Will; Sorenson, Kelly; Stewart, Glenn; Stirling, Susan; Suddjian, David; Sullivan, Judy; Tallerico, Karla; Todd, Nick; Watt, Mary; Weygandt, Clara; Yasuda, Susan; Young, Paul

2007 Two pairs of peregrine falcons continue to reside year round and nest at Morro Rock. The south side pair nested in the 'diving board eyrie'. The east side pair nested again on the wide ledge. Both pairs successfully fledged 3 young the third week of May. Six fledglings and four adults in flight at the Rock, a sight to behold!

Occasional nest watch observers: Judy Sullivan, Bob Isenberg, Cleve Nash, Steve Schubert, Roy Burke

Adult peregrine preening beneath the 'overhang' photo: Mike Baird

In July, a memorial climb in honor of Brian Walton, coordinator of the Santa Cruz Predatory Bird Research Group for 31 years, was made by 5 climbers from SCPBRG. Judy Sullivan wrote about the event:

Yesterday morning Brian Latta and five others made a climb up the Rock to collect egg shell fragments from the south side eyrie. The group who made the climb was made up of colleagues and relatives of Brian Walton, who passed away a month ago at the age of 55. Brian was the head of the Santa Cruz Predatory Bird Research Group for many years. Watching from below was a large gathering of Brian W's relatives, colleagues, and admirers. An informal remembrance of Brian was held right before the ascent, with words being said about the amazing impact he had on the recovery effort of the west coast Peregrine Falcons. It was an emotional day. Brian lived his life well and contributed greatly to all of us. If you would like to honor Brian and his work, consider making a donation to the SCPBRG. Without their work, we wouldn't have these amazing birds to enjoy.

2008 The south side pair nested in the 'diving board eyrie'. The east side pair nested again on the wide ledge. The south side fledged 2 young. The east side pair successfully fledged 3 young.

The south side adult male apparently disappeared several days after the chicks hatched. A new resident adult male aggressively chased the recently fledged young, which then relocated to the east side and joined the three fledglings there into a single group of 5 fledglings! All were cared for by the adoptive parents. The five fledglings - often in flight together or perched in close proximity on Morro Rock, on the sand spit, or on the stacks and roof of the electrical power plant across the bay - eventually dispersed from the area.

Occasional nest watch observers: Judy Sullivan, Bob Isenberg, Cleve Nash, Steve Schubert

2009 The south side pair fledged 3 young in mid-May from the pothole eyrie with the long white streak beneath, to the right (east) of the traditional 'diving board' pothole. The north side pair fledged 2 young – a male and a female- the following week over Memorial Day weekend, from the pothole eyrie last occupied in 2001.

Occasional nest watch observers: Judy Sullivan, Bob Isenberg, Cleve Nash, Steve Schubert, Jeff Sipple

The American Peregrine Falcon (*Falco peregrinus anatum*) was removed from California's list of endangered species in 2009, although it is still a fully protected species by state and federal laws.

2010   The south side pair fledged 2 young in early May from the 'diving board' eyrie. The east (north) side pair fledged 4 young in late May from the eyrie beneath the cap- or bell-shaped rock – the 'dome' eyrie - on the east face.

Occasional nest watch observers: Judy Sullivan, Bob Isenberg, Cleve Nash, Steve Schubert

The 29 minute television program 'Peregrine Falcon – Saved!' aired on a cable channel broadcast, produced and edited by Val Jeffery for Cupertino Television Productions, program #954. Steve Schubert and Bob Isenberg were interviewed at Morro Rock for the program, which is also posted online at **www.YouTube.com**, entitled 'The Better Part – Peregrine Falcon – Saved!'

2011   The north side pair fledged 4 young during the third week of May in the eyrie at the cluster of 'bowling ball' holes. The south side pair failed during the first nesting attempt in the oblong-shaped pothole eyrie (below the 'diving board' pothole), relocating to another pothole eyrie lower on the Rock face and fledged 3 young in the second nesting attempt during the last week of June.

In June, two juvenile peregrines found entrance into one of the Morro Bay power plant towers and were trapped down inside a large duct that connects the tall tower to the main building. The peregrines were captured with a pelican net and released by two volunteer rehabbers with Pacific Wildlife Care, first being given a 30-minute safety briefing and wearing hard hats and safety glasses before climbing into the duct by a ladder.

Occasional nest watch observers: Judy Sullivan, Bob Isenberg, Cleve Nash, Steve Schubert

2012   The south side pair nested at the 'diving board' eyrie. Eggs hatched about April 8th, Easter Sunday. Two male and two female chicks fledged during the third week of May. The north side pair failed at the first nesting attempt at the 'bowling ball' hole. A second nesting attempt produced one 'Solo' chick, fledged from the east side on a ledge near the 'dome' eyrie, during the first week of June.

Occasional nest watch observers: Judy Sullivan, Bob Isenberg, Cleve Nash, Heather O'Connor, Steve Schubert

Pacific Coast Peregrine Watch started a Facebook page with postings at: **www.facebook.com/PacificCoastPeregrineWatch/**

This year a USFWS post-delisting peregrine falcon nesting survey was conducted at a number of sites throughout the state.

2013 The south side pair failed at two nesting attempts, at the 'diving board' eyrie and a pothole below. Three chicks fledged over Memorial Day weekend at the north side 'bowling ball' pothole eyrie.

Occasional nest watch observers: Bob Isenberg, Cleve Nash, Judy Sullivan, Steve Schubert

2014 Chicks hatched approximately April 18th at the south side eyrie, in the pothole with the long white streak - the 'waterfall' eyrie (right [east] of the 'diving board' eyrie). One chick fledged on or about the date of May 22nd. Chicks hatched in April at the north side eyrie, at the 'bowling ball' potholes. Two chicks fledged late May. Surprisingly, the two north side fledglings by early June were residing on the south side and being fed and cared for by that nesting pair, and had joined with the single south side fledgling. A similar event occurred in 2008 when the south side fledglings were 'adopted' by the north side nesting pair.

Occasional nest watch observers: Bob Isenberg, Cleve Nash, Judy Sullivan, Steve Schubert

2015 The south side pair failed in two nesting attempts. The north side pair fledged 3 young over Memorial Day weekend in May - the eyrie was located at the slanted ledge below the bell (dome)-shaped rock on the east side.

The north side adult female was photographed in July by Cleve Nash and identified by her left leg band (R/23). She had been banded in 2012 after fledging from a nest tray with a gravel substrate that had been placed 250-feet high on the catwalk encircling one of the two smokestacks on the Moss Landing Power Plant, overlooking Elkhorn Slough near the edge of Monterey Bay. At 3 years old she was now the north side breeding female at MorroRock!

Occasional nest watch observers: Bob Isenberg, Cleve Nash, Gordon Robb, Steve Schubert

2016 The 20th annual Morro Bay Winter Bird Festival, held each year during the month of January, was attended by several hundred registrants from across the country. The peregrine falcon slide talk and field trip to Morro Rock to observe the falcons has been offered as a workshop every year of this event.

The south side pair was present throughout the year but apparently no eggs were laid. There was speculation the old female may be beyond egg-laying age. The north side pair fledged 3 young on May 24th, at the 'bowling ball' potholes eyrie on the north side.

Occasional nest watch observers: Bob Isenberg, Cleve Nash, Gordon Robb, Steve Schubert

2017 The south side and north side nest sites failed this nesting season. During a period of time in May prey was delivered and chick vocalizations were heard by observers at a north side eyrie, but then were never seen or fledged. This was the first year since 1999 that no wild hatched peregrine falcon chicks successfully fledged from Morro Rock (although 3 chicks were fostered into the south side nest and fledged that year). The south side nesting female – suspected to be the same unbanded falcon to have resided at the Rock since 2001 - would now be about 17 years of age and very old for a wild peregrine falcon, and is perhaps no longer producing viable eggs.

Occasional nest watch observers: Bob Isenberg, Cleve Nash, Gordon Robb, Steve Schubert

The Morro Bay National Estuary Program posted a Morro Rock peregrine falcon article on its website at: **www.mbnep.org/morro-bay-wildlife-spotlight-peregrine-falcon/**

The local publication **Journal Plus**, in the August, 2017 issue, featured an article "Bird Watching On The Central Coast", highlighting Bob Isenberg's many years of volunteer efforts and enthusiasm educating the visiting public about the falcons of Morro Rock. Cover photo - a local wintering peregrine falcon harassing a bald eagle - by Cleve Nash.

Summary

The peregrine falcons at Morro Rock experienced documented nesting failures in 1977 and 1978, when no eggs hatched (there were likely other previous unrecorded nesting failures in the 1950's and '60's as well, following the introduction of the chlorinated hydrocarbon pesticide DDT). A fifteen-year intensive research and management program - including nest site manipulation by removal and captive hatching of thin-shelled eggs, double clutching, banding of young, fostering captive hatched peregrine chicks into the eyrie, and collection and analysis of eggshells and prey remains - was conducted each year at Morro Rock from 1977 through 1992. Nest site surveillance by employed and/or volunteer observers has occurred since 1967 through the present. The falcon pair successfully raised two young in 1993, the first nesting attempt without human intervention and nest manipulation since 1977 - a time span of 16 years - and hatched, reared, and fledged their own young. The pair produced one nestling in 1994, which did not survive, successfully fledged two young in 1995, fledged one young in 1996 and 1997, but were unsuccessful hatching eggs in two nesting attempts in 1998 and again in 1999. In 1999 three young captive hatched chicks were fostered into the nest and fledged successfully. A new pair of adult falcons nested in 2000 and successfully hatched and fledged two young. Amazingly, two nesting pairs of falcons occupied Morro Rock in 2001 but only one of these pairs was successful fledging young, at the north side eyrie. Two nesting pairs were successful fledging young in 2002. In 2003, the south side pair fledged two young but the lone east side female did not nest.

Two nesting pairs successfully fledged young in 2004. In 2005 the south side pair fledged three young but the east side pair failed to produce young after attempting to nest. One young fledged from the south side eyrie in 2006 on the second nesting attempt, but the east side pair again failed to produce young, after two apparent nesting attempts. In 2007, the south side pair fledged three young and the east side pair also fledged three young, a remarkable event in the decades-long history of recorded observations at Morro Rock.

Two peregrine falcon nesting pairs have occupied Morro Rock during the years of observation 2001 through 2017, referred to by observers as the 'south side' and 'north side' – or in some years the 'east side' - pairs. Surprisingly, during two nesting seasons in 2008 and 2014, recent fledglings from one nest site moved around to the adjacent nesting territory and were fed and cared for there by their non-biological 'adoptive' parents, in addition to providing for their own fledged young.

Volunteers continue the yearly monitoring of peregrine falcon nesting activities at Morro Rock, begun more than 50 years ago. Many individuals, including volunteers and professional wildlife biologists, state and federal wildlife agencies, and non-profit conservation organizations have participated in the Morro Rock peregrine falcon nest watch and management program over the years, and their efforts and dedication are greatly appreciated.

During the years 1977 through 1992 (and again in 1999 and 2001), the Santa Cruz Predatory Bird Research Group, its staff and volunteers, conducted an intensive nest management program at Morro Rock. These efforts are summarized as follows: number of years the nest sites were manipulated (17); number of years the nests were double clutched (3); number of eggs hatched in captivity (17); number of young fostered into the eyrie (34); number of fostered peregrines fledged (32)

During this same period of time, the following Morro Rock climbers rappelled to the nest sites, participated in nest site management, removed eggs for captive incubation, leg-banded and fostered young chicks, and collected eggshell fragments and prey remains for analysis: Brian Walton, Ron Walker, Carl Thelander, Merlyn Felton, Lee Aulman, Craig Himmelwright, Matt Nixon, Mitch Siemens, Victor Apanius, Galen Rowell, Charles Porter, Kurt Stolzenburg, Scott Francis, Jamey Eddy, Brian Latta

The once large feral cat population at the Rock preyed upon the songbirds, native rodents, and other wild animals. Abandonment of cats by pet owners and illegal feeding by well-intentioned visitors aggravated the problem. Cats climbing on the Rock are a potential threat to perched and inexperienced fledgling falcons. Adult falcons on more than one occasion in the past have been observed to be very agitated by the presence of cats and have made attacks. A feral cat high up on the Rock was observed killed by a peregrine falcon in 1992. The presence of nesting peregrine falcons and other sensitive wildlife species at Morro Rock, which has been established as an Ecological Reserve for the protection of wildlife and its habitat, should prompt the continued removal of abandoned predatory, non-native feral cats. Personnel from Morro Bay State Park, California Department of Fish and Wildlife, and the City of Morro Bay are encouraged to continue enforcement of no-feeding and abandonment regulations.

Although the falcons are accustomed to boat, vehicle, and pedestrian activity far below, the presence of illegal hikers high on the Rock and near the nest site causes great disturbance. The alarmed peregrines fly at close range and vocalize agitatedly at the intruders. The Ecological Reserve status and climbing restrictions at Morro Rock were established, in part, to prevent frequent nest site disturbances and the potential for abandonment and nesting failure. No-climbing signs to the general public are posted, yet hikers and trespassers have been frequently observed climbing on the steep and dangerous slopes. Aircraft flyovers, and in more recent years the flying of remote controlled drones, occasionally enter the airspace very near Morro Rock. Approaches too close to the Rock and harassment of wildlife may be violations of state and federal laws, as well as local city and state park regulations.

After becoming nearly extinct as a breeding species in California, the research and management program conducted for several decades in conjunction with peregrine falcon recovery efforts across the continent, assisted with the dramatic recovery of a once imperiled species. In 1970, only two recently productive nesting pairs of peregrine falcons were known in California (one was at Morro Rock); now, decades later, there may be more than 300 peregrine falcon nesting pairs throughout the state. Presently, at least seven coastal and four inland peregrine falcon nesting sites are known to occur in San Luis Obispo County on the Central California Coast. Sightings of peregrines continue to increase locally and elsewhere throughout the state due to an increasing falcon population - both year-round residents and wintering migrants that have traveled from afar.

photo: Cleve Nash
Shell Beach peregrine in flight

The removal of the American Peregrine Falcon from the federal Endangered Species List in 1999 was credited to the implementation and success of the federal Endangered Species Act, and due in large part to the combined efforts and collaborations of the U.S.F.W.S. administered Peregrine Falcon Recovery Program and The Peregrine Fund, a non-profit organization. This delisting was viewed by some raptor biologists and conservationists as being perhaps premature and not yet fully warranted. They argued it had not been proved conclusively that the species was successfully reproducing and sustaining its population throughout its historical range, and would require further long-term monitoring.

Following delisting, this monitoring effort was coordinated by the U.S. Fish and Wildlife Service, with assistance in California from SCPBRG staff and volunteers, conducting post-delisting peregrine falcon surveys every 3 years from 2003 to 2015. These surveys monitored territory occupancy, nest success, and productivity. The results of these surveys documented the continued recovery of the peregrine falcon population and increasing number of nesting pairs, a species on the road to recovery.

Public education and awareness, protection of important foraging and nesting habitats, and regulating toxic chemicals that concentrate in ecological food chains have been and continue to play an important role sustaining a viable peregrine falcon population. The remarkable recovery of the peregrine falcon population is a success story that gives cause for hope and recognition of the continuing need to take action for the many other imperiled species of flora and fauna that constitute this continent's rich biodiversity.

Over the years thousands of people - local residents, students and field trip participants, tourists and travelers from all over the world - have witnessed the spectacular aerial flights, hunting forays, courtship and nesting behaviors, and other fascinating day-to-day behaviors in the lives of the peregrine falcons at Morro Rock. There is renewed optimism for a species once on the edge of extinction. Many more intriguing observations and stories about the Morro Rock peregrine falcons remain to be told throughout the years.

Read yearly summaries and updates about the nesting peregrines, find where to look for the falcons at the Rock, and view photographs of the falcons taken by local wildlife photographers, posted at:

www.pacificcoastperegrinewatch.org and

www.facebook.com/PacificCoastPeregrineWatch/

Acknowledgements

Brian Walton, Janet Linthicum, and Brian Latta from the Santa Cruz Predatory Bird Research Group were all very helpful providing information and reviewing the draft copies of this manuscript. Merlyn Felton reviewed the article and provided many details and fascinating stories regarding his experiences for several years serving as a Morro Rock falcon nest guard. Several others were helpful in reviewing and commenting on the article and Tery Drager from SCPBRG was instrumental in its production. Eggshells were measured by Sam Sumida of the Western Foundation of Vertebrate Zoology. Photographs are by Steve Schubert, Cleve Nash, Gary Robertshaw, Mike Baird, and from the MCAS files and the SCPBRG slide archive. John Schmitt prepared the illustrations and long ago introduced me to the "art" of observing peregrine falcons when we worked and camped out as nest site attendants at a peregrine site in San Luis Obispo County, within Los Padres National Forest. The Telegram Tribune and Sun Bulletin newspapers, as well as local KSBY television news station have provided coverage over the years about the nesting falcons at Morro Rock. More recently, social media sites and related websites are appreciated for informing and updating local residents and visitors regarding the local nesting peregrine falcons and recent observations, promoting further interest and awareness.

Telegram Tribune newspaper article, June 3, 1978

Falcon observers at the Rock

Finally, to all those who will continue to observe the falcons at the Rock with awe and wonder for years to come, good watching!

40-YEAR SUMMARY OF PEREGRINE FALCON NESTING AND MANAGEMENT AT MORRO ROCK, CA

YEAR	MANIPULATED	DOUBLE CLUTCHED	MINIMUM # OF EGGS LAID	# OF YOUNG FOSTERED	# OF YOUNG FLEDGED	# EGGS CAPTIVE INCUBATED/HATCHED	% EGGSHELL THINNING**	NOTES
1967	no	—	unknown (2+)	—	1	–/–	—	One egg died in nest?
1968	no	—	unknown (3+)	—	3	–/–	—	One fledgling died
1969	no	—	0?	—	0	–/–	—	Ad. female died at egg laying
1970	no	—	0	—	0	–/–	—	No nesting
1971	no	—	unknown (3+)	—	3	–/–	—	New ad. female
1972	no	—	unknown (2+)	—	0	–/–	—	Two nestlings stolen
1973	no	—	3	—	0?	–/–	—	One ad. disappeared; new ad. male?
1974	no	—	4	—	4	–/–	—	Four young fledged
1975	no	—	3	—	2	–/–	—	One egg disappeared; female disabled after wire strike
1976	no	—	3	—	2	–/–	10.4	New ad. female; one egg addled
1977	yes	no	unknown	2	1	0/0	15.9	2 fostered chicks; ad. male shot; one nestling died in nest
1978	yes	no	3	1	1	0/0	15.4	New ad. male; one egg disappeared; two eggs addled
1979	yes	yes	6	2	2	6/1	(1) 21.7 (2) 69.9	Two sets of eggs removed
1980	yes	yes*	7	2	2	4/1	(1) 27.2 (2) 29.9	Broken 1st clutch; two eggs of 2nd clutch addled
1981	yes	yes	8	2	2	7/4	(1) 28.3 (2) 28.7	Four eggs addled
1982	yes	yes	7	2	2	6/5	(1) 25.3 (2) 26.1	Two eggs addled
1983	yes	no	4	3	2	3/0	24.2	Four eggs addled; one nestling died
1984	yes	yes*	6	2	2	3/2	(1) 20.9 (2) 26.4	Broke most of 1st clutch; two eggs addled
1985	yes	no	3	2	2	2/0	18.4	Two thin eggs unviable
1986	yes	no	3	2	2	3/1	24.5	One egg captive hatched; two fostered chicks fledged
1987	yes	no	4	2	2	1/0	22.0	New adults; three eggs broken; one DOA
1988	yes	no	5	2	2	4/2	(1) 10.2 (2) 24.2	Ad. male died after wire collision
1989	yes	no	3	2	2	2/0	25.3	New ad. male; two eggs removed
1990	yes	no	4	1	1	3/1	23.6	One egg captive hatched; one fostered chick fledged
1991	yes	yes?*	unknown (1+)	2	2	1/0	15.4	Broken 1st clutch; one egg in 2nd clutch DOA
1992	yes	no	4	2	2	4/0	23.6	New ad. female; all four eggs DOA

*naturally
(1) 1st clutch
(2) 2nd clutch

YEAR	MANIPULATED	DOUBLE-CLUTCHED	MINIMUM # OF EGGS LAID	# OF YOUNG FOSTERED	# OF YOUNG FLEDGED	# EGGS CAPTIVE INCUBATED/ HATCHED	% EGGSHELL THINNING**	NOTES
1993	no	no	2	–	2	–/–	23.6	Fledged their own young (2)
1994	no	yes*	unknown (1+)	–	0	–/–	(1) 19.5 (2) 19.5	Lost first clutch; nestling disappeared
1995	no	no	2	–	2	–/–	22.6	Fledged their own young (2)
1996	no	no	2	–	2	–/–	22.5	Fledged their own young (1)
1997	no	no	1	–	1	–/–	22.7	Fledged their own young (1)
1998	no	yes*	unknown (2+)	–	0	–/–	(1) 13.9 (2) 21.2	Unhatched eggs in two nesting attempts; one egg broken; one addled
1999	yes	yes*	unknown (1+)	3	3	–/–	?	Unhatched eggs; three fostered chicks fledged; ad. female died
2000	no	no	3	–	2	–/–	?	New ad. male and female; fledged their own young (2)
2001	no	no	s-side nest unknown n-side nest 3	–	s-side 0 n-side 3	–/–	?	Two pairs of falcons nested; s-side failed; n-side fledged 3
2002	no	no	s-side nest 1 n-side nest 3	–	s-side 1 n-side 3	–/–		Two pairs successful; s-side fledged 1; n-side fledged 3
2003	no	no	s-side nest 2	–	s-side 2	–/–		One new ad. male; 1 pair nested; one female on e-side; s-side fledged 2
2004	no	no	s-side nest 2 e-side nest 3	–	s-side 2 e-side 3	–/–		Two pairs successful; s-side fledged 2; e-side fledged 3
2005	no	no	s-side nest 3 e-side nest unknown	–	s-side 3 e-side 0	–/–		Two pairs nested; s-side fledged 3; e-side failed
2006	no	yes* (both pairs)	s-side nest unknown (1+) n-side nest unknown (1+)	–	s-side 1 n-side 0	–/–		Both pairs failed first nesting attempt; s-side fledged 1; n-side failed
2007	no	no	s-side nest 3 e-side nest 3	–	s-side 3 e-side 3	–/–	?	Two pairs successful; s-side fledged 3; e-side fledged 3

*naturally

(1) 1st clutch
(2) 2nd clutch

Totals: 40 years of nesting records	MINIMUM # OF EGGS LAID	# OF YOUNG FOSTERED	# OF YOUNG FLEDGED	# EGGS CAPTIVE INCUBATED/ HATCHED	% EGGSHELL THINNING**	
	126+	34	79	49/17	21.4% average	

www.ingramcontent.com/pod-product-compliance
Lightning Source LLC
Chambersburg PA
CBHW051928210526
45473CB00006B/2172